AIRBUS A300 and 310

P. R. SMITH

Copyright © Jane's Publishing Company Limited 1987
First published in the United Kingdom in 1987 by
Jane's Publishing Company Limited
238 City Road, London EC1V 2PU
in conjunction with DPR Marketing and Sales
37 Heath Road, Twickenham, Middlesex TW1 4AW

ISBN 0 7106 0428 9

Printed in the United Kingdom by Netherwood Dalton & Co Ltd

JANE'S TRANSPORT PRESS

Cover illustrations

Front: **Lufthansa — German Airlines** (LH)
Lufthansa, the West German national airline, operates scheduled jet services over a vast intercontinental route system that connects nine domestic and 118 foreign points. The airline operates a large fleet of aircraft that includes the Airbus A310. The aircraft are used on internal and European routes, the latter including Frankfurt-London (Heathrow). *(Airbus Industrie)*

Rear: **Pan Am World Airways** (PA)
Pan Am is one of the world's major airline companies with a route network that extends over 144 837 km (90 000 miles) and provides connections to most of the world's capital cities. The carrier was formed on 14 March 1927, and an inaugural flight was made over a Key West, Florida-Havana, Cuba route seven months later. The carrier has its headquarters in New York and bases its aircraft at New York (JFK), Miami, Los Angeles, San Francisco, London (Heathrow), Frankfurt, West Berlin and Tokyo. An A300B4 is seen on its delivery flight. *(Airbus Industrie)*

Right: **Saudia — Saudi Arabian Airlines** (SV)
Saudia, the biggest airline in the Arab world, maintains scheduled jet services to over 20 Saudi Arabian domestic airports, and flies to over 90 foreign destinations in the Middle East, Africa, southern & eastern Asia, Europe and the USA. The company was formed in 1945, and utilised DC-3 aircraft. Today the company flies Boeing 707, 737, 747, L-1011 and Airbus A300-600 aircraft. *(Airbus Industrie)*

Introduction

Airbus Industrie was set up as a 'Groupement d'Intérêts Economiques' to manage the development, manufacture and marketing of the twin-engined, large-capacity, short/medium range A300B passenger aircraft. Members of the conglomerate company are Aérospatiale SNI (France), Deutsche Airbus GmbH (West Germany), British Aerospace (United Kingdom), Construcciones Aeronauticas SA (Spain), with Fokker NV (Netherlands) and Belairbus (Belgium) being associate members.

Construction of the first A300B, a B1, commenced in September 1969. This aircraft, F-OCAZ, made its inaugural flight on 28 October 1972, and was followed by the second B1 a little under five months later. Following 1580 hours of test flights, French and West German certification was granted in March 1974, with the FAA clearance two months later. The first A300B2 for commercial service entered operations with Air France on 23 May 1974 over the Paris (Orly)-London (Heathrow) route.

The prototype aircraft was equipped with two CF6-50A engines and was 2.65 metres (8 ft 8.75 in) shorter than the production models. The A300 has been built in a total of seven different versions: the B1, a prototype version, of which one remains in service (with TEA of Belgium); the B2, which was the basic production version; and the B4, which was a long-range development of the B2, with the same capacity but increased design weights and fuel capacity. The A300-600, which is an advanced version of the B4-200, first flew on 8 July 1983. The aircraft has an increased overall weight, greater passenger capacity, and is 0.46 metres (1 ft 4 in) longer than the B4-200 version. The -600 also has more fuel-efficient engines, incorporating either the General Electric CF6-80C2-A, Pratt & Whitney JT9D-7R4H1, PW4156, PW4058, or Rolls-Royce RB.211-524D4A types. A -600ER version with an extended range is also available. Wingtip fences are added for greater fuel efficiency, together with an additional fuel/trim tank in the tailplane, carbon brakes and greater use of composites. An A300C, convertible, and A300F, freighter, are available for every mark of the A300 range.

The A310 was launched in July 1978. In comparison with the B2 and B4 variants of the A300, the 310 is shorter by 13 frames on the overall length. The aircraft also has new, advanced technology wings of reduced span and area; new and smaller horizontal tail surfaces; common pylons able to support all types of General Electric and Pratt & Whitney engines offered; and landing gear modified to cater to these changes in size and weight. It was also the first model to feature Airbus Industrie's advanced two-man FFCC (Forward Facing Cockpit Control) flight deck. The prototype A310 first flew on 3 April 1982 and entered service with Lufthansa on 12 April 1983. The aircraft is certified with General Electric CF6-80A3 or CF6-80C2, Pratt & Whitney JT9D-7R4D1, or -7R4E1 engines. From mid-1987, the PW4000 will also be available. The A310 is available in both convertible and freighter versions as well as the -200 and -300 variants: the former is the basic passenger aircraft, with the -300 series being a long-range type, with an increase in take-off weight, greater fuel capacity, the addition of wing fences for greater fuel economy, as well as many other extras.

I would like to extend my thanks to everyone who has contributed to this book, as well as those who have helped typecheck the copy. My sincere thanks must also go to Nicholas Moll, who designed the layouts of this and Air Portfolio 1: *Boeing 737,* and Air Portfolio 2: *Shorts 330 and 360.* His superb efforts have been greatly appreciated. I would like to dedicate this book to my father, whose help and understanding in the preparation of it has been greatly appreciated.

TABLE OF COMPARISONS

	A300B2	A300B4	A300-600
Max. accommodation	345	345	375
Wing span	44.84 m (147 ft 1 in)	44.84 m (147 ft 1 in)	44.84 m (147 ft 1 in)
Length	53.62 m (175 ft 11 in)	53.62 m (175 ft 11 in)	54.08 m (177 ft 5 in)
Height	16.53 m (54 ft 2.75 in)	16.53 m (54 ft 2.75 in)	16.62 m (54 ft 6.5 in)
Max. t/o weight	137 000 kg (302 000 lb)	150 000 kg (330 700 lb)	165 000 kg (363 765 lb)
Max. cruis. speed	937 km/h (582 mph)	937 km/h (582 mph)	897 km/h (557 mph)
Maximum range	3700 km (2300 miles)	5190 km (3225 miles)	6912 km (4295 miles)
Service ceiling	10 700 m (35 000 ft)	10 700 m (35 000 ft)	12 200 m (40 000 ft)
	A310-200	**A310-300**	
Max. accommodation	280	280	
Wing span	43.90 m (144 ft 0 in)	43.90 m (144 ft 0 in)	
Length	46.66 m (153 ft 1 in)	46.66 m (153 ft 1 in)	
Height	15.80 m (51 ft 10 in)	15.80 m (51 ft 10 in)	
Max. t/o weight	142 000 kg (313 055 lb)	153 000 kg (337 305 lb)	
Max. cruis. speed	667 km/h (414 mph)	667 km/h (414 mph)	
Maximum range	7042 km (4376 miles)	9266 km (5757 miles)	
Service ceiling	11 275 m (37 000 ft)	11 275 m (37 000 ft)	

Previous page: **Kuwait Airways Corporation** (KU)

Kuwait Airways can trace its history as far back as March 1954, when Kuwait National Airways Company began services over a Kuwait-Basrah line, using DC-3 aircraft. By the middle of the following year, the Kuwaiti government had gained a 50 per cent holding in the carrier which became known as Kuwait Airways Corporation (KAC). In March 1964, KAC inaugurated Comet jet flights to Europe (London Heathrow, Paris, Frankfurt and Geneva). Today the company flies to points within the USA, Europe, southern and eastern Asia, North Africa and the Middle East. Kuwait Airways operates a modern fleet of Boeing 727, 747, Gulfstream 3, BAe 125, A300C-600 and A310-200 aircraft. An example of the latter type can be seen here just after take-off. *(Airbus Industrie)*

Below: **Aerovias Condor de Colombia (Aerocondor)** (OD)

Aerocondor, a privately owned carrier, was formed in February 1955. Scheduled domestic passenger and cargo services were operated domestically between Bogotá, Baranquilla, Cali, Cartagena, Santa Marta, Medellín, San Andreas, Riohacha and Cúcuta. International services were operated to Miami, Guatemala City, Santo Domingo, Panama City, Puerto Principe, Aruba and Curaçao. At the time of the airline's demise in 1981, the fleet consisted of Boeing 707, 720, 737 and Lockheed Electra aircraft. Aerocondor operated this Airbus A300 between 1977 and 1979. *(P Hornfeck)*

Opposite: **Air Afrique** (RK)

Air Afrique is a multinational African airline that serves as the international flag carrier for ten states: Benin, Burkina Faso (originally Upper Volta), Central African Republic, Chad, Congo Republic, Ivory Coast, Mauritania, Niger, Senegal and Togo. The carrier was formed in June 1961, with ownership being held by member national governments who own 72 per cent, and Sodetraf, a French consortium, controlled by UTA French Airlines (28 per cent). Air Afrique operates to points within the USA, Europe, Saudi Arabia and to various African regional countries. The airline operates a modern fleet of DC-10, Boeing 727, 747 and Airbus A300 aircraft. Routes operated by the latter type include: Dakar-Nouakchott-Niamey-Ndjamena-Jeddah; Abidjan-Niamey-Marseilles-Paris (CDG); and Dakar-Conakry-Abidjan-Cotonou-Douala-Brazzaville. *(Airbus Industrie)*

Air Algérie (AH)

Air Algérie was created in 1947 as Compagnie Générale de Transports Aériens. In June 1953 a merger with Compagnie Air Transport formed Air Algérie. The company became Algeria's national flag carrier on 10 April 1963, when the government acquired a 51 per cent shareholding. Today the company's scheduled operations connect the major cities of Algeria with over 35 foreign destinations across North, West and Central Africa, the Middle East, and Europe. Domestic Algerian scheduled air taxi and agricultural services are maintained by an affiliate carrier, Inter Air Services, using Air Algérie equipment. Air Algérie operates a fleet of Boeing 727, 737, Super Hercules, and A310-200 aircraft. The latter type can be seen plying the routes between West Africa and the Middle East from its base in Algeria. *(Airbus Industrie)*

Air France (AF)

Air France provides scheduled services over one of the world's largest and best established intercontinental route networks. The company was formed at the end of August 1933 by a merger of four French airlines. Until September 1939, when the Second World War halted all activities, Air France quickly established itself as a leading world airline, operating 90 aircraft of 11 types. The company was nationalised in June 1945 and took over the entire French air system in January 1946. Today the carrier serves over 130 foreign destinations as well as several domestic trunk routes. In 1976, Air France began supersonic Concorde flights. The company operates a mixed fleet of aircraft, including the A310 (of which it was a launch customer) and A300. The carrier has ordered the 'all-new' A320 for delivery from the end of the 1980s. A310 aircraft are widely utilised and are placed on such routes as Paris (CDG)-London (Heathrow), which is one of the busiest in the world. F-GEMA, an A310, is seen here in flight during one of its many sectors. *(Airbus Industrie)*

7

Air-India (AI)

Air-India, the government-controlled international airline, operates scheduled jet services to over 50 destinations in Europe, the Middle East, Africa, the Indian Ocean, southern and eastern Asia, as well as to Australia and North America. The airline operates a fleet of Boeing 707, Boeing 747, Airbus A300 and A310 aircraft. An example of the Airbus A300 is seen here at Bombay, awaiting passengers.
(J J Wadia)

Air Inter (IT)

Air Inter, the primary domestic airline of France, was formed in November 1954, and commenced operations over a Paris-Toulouse route in June 1960. The company shares are held by Air France (25 per cent), SNCF French Railways (25 per cent), UTA French Airlines (14.7 per cent) and various other transport companies, commercial groups, banks and Chambers of Commerce. The airline operates more than 400 daily flights over a route network that connects the most important cities and resort areas on the French mainland and in Corsica. In terms of passenger traffic, Air Inter ranks among the top ten in western Europe. The airline operates a fleet of F-27, Caravelle, Dassault Mercure (of which it is the world's only operator) and A300 aircraft, one of which is shown taking off from a regional airport. Air Inter has an order for 20 A320 types for delivery from 1988.
(Airbus Industrie)

9

Air Jamaica (JM)

Air Jamaica, the national flag carrier of Jamaica, and one of the leading operators in the Caribbean, was established in late 1968 by the Jamaican government and Air Canada. The company's inaugural services were operated from Kingston to New York, and Kingston to Miami. The airline provides jet services to the USA, Canada, the Cayman Islands, Puerto Rico, and Haiti. Air Jamaica operates a fleet of Boeing 727 and Airbus A300B4 aircraft. An example of the latter type can be seen here arriving at Miami. *(A J Mercer)*

Air Niugini (PX)

Air Niugini, the national airline of Papua New Guinea, began operations in November 1973, with the company becoming the flag carrier on 16 September 1975, when Papua New Guinea gained its independence from Australia. Scheduled flights are maintained on international routes to Australia, Hong Kong, Indonesia, New Zealand, the Philip-pines, Singapore and the Solomon Islands. Domestic scheduled services are also maintained throughout Papua New Guinea. In 1985 an A300 was added to the fleet and can be seen operating over many international sectors, bearing what must be one of the most colourful of Airbus liveries. *(DPR Marketing and Sales)*

Below: **Air Seychelles** (HM)

Air Seychelles maintains scheduled and charter inter-island passenger services within the Seychelles, and also operates intercontinental flights to Europe. The wholly government-owned airline was established in July 1979, when the Seychelles government acquired and combined Air Mahe and Inter Island Airways. Air Seychelles operates Britten-Norman Trislanders and one Airbus A300B4, which is leased from Air France. *(K. Wright)*

Opposite: **Alitalia** (AZ)

Alitalia, the national carrier of Italy, was established in September 1946 and began flight operations the following year over a Turin-Rome-Catania route. At that time the company was known as Aerolinee Italiane Internazionali. It was not until 1957, through a merger with LAI Italian Airlines, that the present name was adopted. A major subsidiary of Alitalia is the premier domestic Italian carrier Aero Trasporti Italiani (ATI). In addition the company also owns real estate, insurance, hotel and investment companies. Alitalia today operates to over 70 destinations in Italy, the rest of Europe, Africa, the Middle East, southern and eastern Asia, and Australia, as well as to North and South America. A fleet of Boeing 747, DC-9, MD-80 and Airbus A300B4 aircraft is operated. An A300 is seen here on arrival at Heathrow from its base at Rome (Fiumicino). *(J Page)*

Above: **Australian Airlines** (TN)

Australian Airlines (known until 1985 as Trans-Australia Airlines) was formed in 1945. Its inaugural flight was made between Melbourne and Sydney the following year, with Douglas DC-3 aircraft being used over this route. In 1949 the domestic routes of Qantas Airways were acquired. Today the carrier is one of the two primary Australian domestic airlines and maintains a comprehensive scheduled network of routes that covers 32 points nationwide, as well as to New Zealand. Australian Airlines is fully government-owned and is run by an appointed Australian National Airlines Commission. The company owns a mixed fleet of Boeing 727, 737, DC-9, BAe 31 Jetstream, and Airbus A300B4 aircraft. One of the latter can be seen here in the company's old livery, landing after a typical domestic flight. *(Airbus Industrie)*

Opposite: **Balair** (BB)

Balair, Switzerland's largest charter airline, operates worldwide passenger and freight flights with a modern fleet of MD-80, DC-10-30 and A310-300 aircraft. The carrier was formed in 1953 and commenced flight operations four years later. Ownership of Balair is made up of Swissair, which owns a 57 per cent holding, other private interests (31 per cent) and the Swiss government (12 per cent). The majority of the airline's business involves the transport of tour groups over intra-European and intercontinental routes, to and from the Swiss gateway cities of Basel, Geneva and Zürich. Regular inclusive and general tour charters serve main destinations in Switzerland, Greece, Spain, Portugal, Turkey, Italy, Cyprus, France, UK, Morocco, Tunisia, the Canary Islands, Kenya, Togo, Gambia, Sri Lanka. the Maldives and the USA. *(Airbus Industrie)*

British Caledonian Airways (BR)

British Caledonian, the UK's first operator of the A310, is Britain's second largest international airline. From its base at London (Gatwick), the carrier maintains scheduled jet flights over a substantial route system that covers Britain and Western Europe, Africa, the Middle East, the Far East and the USA. The company was formed in 1970 through a merger of British United Airways (BUA) and Caledonian Airways. Until 1985 BCAL operated a successful daily 'Gatwick-Heathrow Airlink' utilising a Sikorsky S-61N. The current fleet includes Boeing 747, DC-10 and BAe One-Eleven aircraft. The two A310 aircraft that the airline used to operate have now been sold. They were used on such lucrative routes as London (Gatwick)-Tripoli (Libya) and London (Gatwick)-Lagos. Just prior to their sale, the Airbus types could be seen operating flights to Paris (CDG). *(Airbus Industrie)*

CAAC — General Administration of Civil Aviation in China (CA)

CAAC is the government-controlled national airline of the People's Republic of China. The carrier was formed in 1962 and succeeded the Civil Aviation Administration of China. In addition to its airline operations, CAAC controls all civilian air transport activities in China (including passenger, cargo, airport, training, agricultural and other specialised services). The carrier maintains scheduled flights over an extensive domestic network and serves international routes in Asia and to the Middle East, Africa, Europe, USA and Australia. CAAC operates a mixed fleet that comprises Boeing, McDonnell Douglas, Ilyushin, Antonov, de Havilland Canada, Lisunov, British Aerospace, Bölkow, Bell, Yanshu, Tupolev, Sikorsky and Airbus types. B-2301, an A310-200, is seen here arriving at Tokyo (Narita) after a flight from Beijing, one of the major sectors for which the Airbus type is used. *(A Okinda)*

Opposite: **China Airlines** (CI)
China Airlines, the privately-owned Taiwanese flag carrier, was established in December 1959 by retired Chinese Air Force personnel. During the early 1960s, the company concentrated on charter services and the rapid development of a domestic network. It was not until six years later that CAL opened an international route between Taipei and Saigon. In 1967 further overseas lines were inaugurated to Japan and Hong Kong. The airline serves destinations in eastern and southern Asia, the Middle East, Europe and the USA. The company operates a modern fleet of Boeing 737, 747, 767, and Airbus A300B4 aircraft types. The latter type operates services to destinations including Hong Kong, Tokyo, Bangkok and Manila. An A300 is seen here on its delivery flight to Taipei.
(Airbus Industrie)

Below: **Condor Flugdienst GmbH** (DF)
Condor, the wholly-owned charter subsidiary of Lufthansa, operates regular inclusive tour services from most West German centres, including Frankfurt, Munich, Stuttgart, Köln/Bonn, Düsseldorf, Hamburg and Hannover. Airbus A300 aircraft are no longer operated on a long-term basis, although one example leased from Hapag-Lloyd has regularly added capacity during the peak summer season. Airbus A310 services link the Federal Republic with Mediterranean 'sunspots', the Far East, and Africa.
(Udo and Birgit Schaefer Collection)

Continental Airlines (CO)

Continental Airlines provides scheduled low-fare jet services to almost 50 domestic destinations in the USA, and to foreign airports in the Central and South Pacific, East Asia, Mexico, Canada and the UK. The company was formed in 1934 when Varney Special Lines began services. In 1936, following the purchase of a Denver-Pueblo route from Wyoming Air Service, the company moved its base from El Paso to Denver, and changed its name to Continental Airlines. In October 1982 the company merged with Texas International Airlines. The airline has a fleet of A300B4, DC-9, DC-10, Boeing 727 and 737-300, as well as MD-80 aircraft. *(N Mills)*

Cruzeiro — Brazilian Airlines (SC)

Cruzeiro, the third largest airline in Brazil, was formed in December 1927 as Kondor Syndikat. Initial services linked Rio de Janeiro, Porto Alegre and other southern Brazilian points. In 1935 international routes were opened to Buenos Aires and San Diego. During the Second World War the carrier was restructured. Douglas DC-3 aircraft were purchased and a name change took place, the company becoming Serviços Aéreos Cruzeiro do Sul. Through the 1940s and 50s existing routes were fortified and new routes added. Two local airlines, SA Viacão Aérea Gaucha (SAVAG) and Transportes Aéreas Catarinense (TAC) were absorbed. The company operates to 26 destinations within Brazil as well as serving foreign cities in Argentina, Barbados, Bolivia, French Guinea, Peru, Surinam, Trinidad and Uruguay. Cruzeiro has operated the A300B4 since 1980 and also utilises a fleet of Boeing 727 and 737 aircraft.
(Udo and Birgit Schaefer Collection)

Opposite: **Cyprus Airways** (CY)
Cyprus Airways operates scheduled flights between Cyprus and over 20 cities in Europe, North Africa and the Middle East. Substantial tour charter operations are also undertaken, some of which are organised by a subsidiary known as Cyprair Holdings. The airline was formed in September 1947 by British European Airways, the Cyprus government and local interests. The following year the airline acquired three DC-3 aircraft and opened routes between Nicosia and Beirut, Cairo, Haifa, Istanbul, Rome and Athens. Today the company operates a fleet of Boeing 707, BAe One-Eleven and Airbus A310 aircraft. The latter type can be seen operating from Larnaca to points such as London (Heathrow) and Frankfurt. Cyprus Airways also has orders for the 'all-new' A320 for delivery from 1989 onwards. *(Airbus Industrie)*

Above: **Dan-Air** (DA)
Dan-Air was established as a charter operator in May 1953 with a base at Southend. The airline is a wholly-owned subsidiary of Davies and Newman Holdings Ltd, shipping agents and brokers. Dan-Air has its main flight base at London (Gatwick) Airport and has maintenance centres at both Man- chester and Lasham. The company serves many UK destinations as well as charter and contract flying to Europe and North Africa. Regional European holiday tour destinations are served from many parts of the UK as well as West Berlin. The airline operates a fleet of BAe One-Eleven, 146, 748, Boeing 727, 737 and Airbus A300B4 aircraft. The latter type is leased from the West German carrier Hapag-Lloyd and is registered G-BMNA. In November 1986 this aircraft was returned to the airline and another, G-BMNB, was purchased outright. *(Udo and Birgit Schaefer Collection)*

23

Eastern Airlines (EA)

Eastern Airlines was established in July 1923 by Pitcairn Aviation as an air mail carrier. The company formed Eastern Air Transport and in August of the following year began scheduled passenger services along a New York-Newark-Camden-Baltimore-Washington-Richmond route. The carrier flies all over the USA and Canada, as well as to Central and South America, the Bahamas, the Caribbean and Bermuda. To enhance traffic at major hub airports, Eastern has marketing agreements with various regional and commuter airlines, providing improved flight connections to and from many small cities and towns. Eastern operates a fleet of DC-9, DC-10, Boeing 727, 757, L-1011 and Airbus A300B4 aircraft. (A J Mercer)

EgyptAir (MS)

EgyptAir was established in 1932 as Misr Airwork, with a change of name to Misrair in 1949. In 1960 the airline once again changed its title, this time to United Arab Airlines. It was not, however, until 1971 that the present name was adopted. EgyptAir operates its eight A300B4s over one of the toughest and most demanding route networks anywhere in the world, maintaining an average of 6.5 hours per day from each aircraft. The airline operates scheduled flights over domestic routes within Egypt, as well as international services from its base at Cairo, to the Middle East, southern and eastern Asia and also to Europe. EgyptAir's initial A300s were leased from the West German charter airline, Hapag-Lloyd. A300B4, SU-BCC *Nout,* is seen here taxiing at Geneva airport, having recently arrived from Cairo. *(Udo and Birgit Schaefer Collection)*

Garuda Indonesian Airways (GA)

Garuda, the government-controlled national airline of Indonesia, was formed in March 1950 to succeed Indonesian Airways. The company inaugurated international services to Singapore, Bangkok and Manila in 1954, and in 1978 took over management control of the separately run Merpati Nursantara Airlines. The airline operates from a main hub at Jakarta to points within Europe, the Middle East, Australia, southern and eastern Asia. Garuda operates a fleet of Boeing 747, DC-9, DC-10, F-28 and Airbus A300 aircraft. The latter type operates several daily non-stop flights between Jakarta and Singapore as well as routes to Hong Kong, Tokyo and Taipei, in addition to domestic trunk services. An A300 is seen here leaving its home base on yet another sector. *(Airbus Industrie)*

Hapag-Lloyd Flug (HF)

Hapag-Lloyd Flug, West Germany's second largest charter carrier, operates an extensive international passenger jet service, plus worldwide cargo flights. The company is a subsidiary of the Hapag-Lloyd shipping organisation and maintains a flight hub at its base in Hannover. The airline inaugurated services in March 1973, and in 1978 Hapag-Lloyd gained financial control of Bavaria Germanair. The majority of the company's work involves the trans-portation of tour-group passengers between West Germany and the main holiday spots of southern Europe, the Canary Islands, North Africa, East Africa and southern Asia. Hapag-Lloyd's fleet consists of Boeing 727, 737 and Airbus A300B4 and C4 types. Using the Airbus aircraft, the carrier now undertakes passenger and freight charters between Europe and both Africa and North America.
(Airbus Industrie)

Above: **Iberia Airlines** (IB)

Iberia, the Spanish government-controlled national airline, is one of the largest passenger carriers in western Europe. The company was formed in June 1927, with an inaugural flight between Madrid and Barcelona in December of that year. Iberia operates to over 20 points on the Spanish mainland, as well as to the Balearic Islands, Melilla and the Canary Islands. International destinations include North America, Europe, the Caribbean, Central and South America, Africa, the Middle and Far East. The carrier's investments include a minority shareholding in Aviaco, Spain's second largest airline. Iberia operates a fleet of Boeing 727, 747, DC-9, DC-10, and Airbus A300B4 aircraft. The latter type can be found plying the routes between Madrid-London (Heathrow), Madrid-Brussels and Malaga-London (Heathrow). *(Airbus Industrie)*

Opposite: **Indian Airlines** (IC)

Indian Airlines was established in June 1953 following the nationalisation of private air carriers, including Air-India and Kalinga Airlines. The government-owned airline was granted domestic and regional international authority, while Air-India gained the primary international services. The company schedules over 200 daily flights along an extensive route system connecting more than 65 cities and towns throughout India, Afghanistan, Bangladesh, Maldive Islands, Nepal, Pakistan and Sri Lanka. Indian Airlines operates a fleet of Boeing 737, BAe 748 and Airbus A300 aircraft. The latter can be seen flying trunk routes from bases in Bombay and Delhi to destinations that include Delhi, Calcutta, Madras, Srinagar and Hyderabad. The airline has also placed orders for the new single-aisle A320. *(Airbus Industrie)*

Iran Air (IA)

Iran Air, the government-controlled Iranian national airline, maintains scheduled jet services within Iran and to 17 foreign destinations in the Middle East, southern and eastern Asia, and Europe. The carrier was formed in 1962 through a merger of Iranian Airlines and Persian Air Services. Since the rise to power of the Ayatollah Khomeini in 1979, the operations of Iran Air have been dramatically curtailed. Before the company operated to 29 destinations, ranging as far east as Tokyo and as far west as New York, and scheduled over 100 weekly departures. As at 1986 less than 30 weekly scheduled flights depart for Iran Air's overseas destinations. A fleet of Boeing 707, 727, 737, 747 and A300 aircraft is utilised. *(Airbus Industrie)*

Kenya Airways (KQ)

Kenya Airways is the government-controlled Kenyan national airline. The company was formed in January 1977 after the sudden demise of East African Airways. Following its inaugural flight between Nairobi and London (Heathrow) a month later, Kenya Airways quickly developed a substantial intercontinental route system. The carrier maintains scheduled flights between major points within Kenya, as well as over international routes covering Africa, southern Asia, the Middle East and Europe. A Kenya Airways subsidiary, Kenya Flamingo Airways, which was a charter company, has been inactive for several years. The parent company operates a fleet of Boeing 707, 720, DC-9, F-27 and the all-new A310-300 aircraft, of which the latter operate the prestige services to London (Heathrow), Paris and Frankfurt. *(Airbus Industrie)*

Above: KLM — Royal Dutch Airlines (KL)
KLM, the world's oldest airline, was founded on 7 October 1919, with an inaugural flight between Amsterdam and London some seven months later. In 1929 Amsterdam-Jakarta flights commenced over the world's then longest air route. Today KLM maintains a comprehensive route network to well over 100 destinations, connecting Amsterdam with the major cities in Europe, the Middle East, southern and eastern Asia, Australia, North, Central and South America, and the Caribbean. The carrier is owned by the Dutch government (55.4 per cent share) and public investors (44.6 per cent share). KLM wholly-owned subsidiaries include KLM Aerocarto, KLM Helicopters and NLM CityHopper. The Airbus A310 was introduced to the airline's fleet to replace DC-8s and supplement DC-9s on short and medium range sectors. Typical routes flown by the Airbus type include Amsterdam-London (Heathrow), Amsterdam-Cairo, and Amsterdam-Paris (CDG). This A310 is seen just prior to its delivery, as it still wears its French test registration. *(Airbus Industrie)*

Opposite: **Korean Air** (KE)
Korean Air was formed on 20 July 1984 following a change of name from Korean Air Lines. A new corporate image and livery was also introduced, with the original 'Swan' symbol being replaced with the 'T'aeguk' which symbolises the Republic of Korea. The carrier has quickly become one of the most modern airlines in the world. The Korean fleet comprises Boeing 707, 727, 747, MD-82, C-212, F-28, DC-10 and A300 type aircraft. The A300F4, a cargo/passenger variant, was introduced to the fleet on 2 August 1986, with a second aircraft arriving a week later. The new aircraft is used on regional and domestic flights. Korean Air operates from its base at Seoul in South Korea to destinations in Europe, the Middle East, Far East, North Africa and the USA. *(Airbus Industrie)*

Laker Airways (GK)

Laker Airways was formed in March 1966 to operate contract inclusive tour and ad-hoc charter services. Operations began later that year. The airline, which was wholly owned by Sir Freddie Laker, had interests in Caribbean Airways and Gatwick Handling. In September 1977 Laker pioneered the art of 'cheap travel' with 'Skytrain' flights from London (Gatwick) to New York, and a year later to Los Angeles. Unfortunately in February 1982 the company ceased operations. At that time, the fleet consisted of BAe One-Eleven, DC-10 and Airbus A300B4 aircraft. Laker had the distinction of becoming the UK's first Airbus operator and, had the company survived, would also have been the UK's first user of the A310 as well. *(A Clancey)*

Luxair — Luxembourg Airlines (LG)

The national airline of the Grand Duchy of Luxembourg was formed early in 1962 and made its inaugural flight over the Luxembourg-Paris line on 2 April of that year. Today the carrier flies scheduled and regular tour services in Europe and the Canary Islands, as well as to North Africa. The company is owned by the government (25 per cent), Luxembourg Steel Industries (15 per cent), the Luxembourg Broadcasting Company RTL (15 per cent), and three Luxembourg banks (International, General and State Savings Bank, each with a 15 per cent share). In addition to scheduled services, Luxair offers executive charters through a Luxair Executive subsidiary. An Airbus A300B4 was purchased to replace the noisy and fuel-thirsty Boeing 707s, which operated over the Luxembourg-Johannesburg route. The Airbus type is operated by Luxavia, a tour company that regularly undertakes international flights. LX-LGP is seen here at London (Heathrow) on an occasional service from its home base. *(R H Vandervord)*

Below: **Martinair Holland NV** (MP)

Martinair Holland is one of Europe's leading charter airlines. The privately-owned Dutch carrier operates worldwide passenger and freight charter and contract services from a base at Amsterdam's Schiphol Airport. The airline was established in 1958 as Martin's Air Charter and began flight operations with a DC-3. Ownership control is held by the Royal Nedlloyd Group (49 per cent share), KLM-Royal Dutch Airlines (25 per cent share) and several banking and investment groups make up the balance. The Martinair Holland business group also includes companies involved in food and catering services, advertising, flight training and the sales and maintenance of business and sport aircraft. The major part of the airline's passenger work involves charter activity in Europe, North and East Africa, South and eastern Asia, and North America. Martinair operates two A310s as well as a fleet of Douglas DC-10 and McDonnell Douglas MD-82 aircraft. The company has two Boeing 747-21ACs on order. *(Udo and Birgit Schaefer Collection)*

Opposite: **Nigeria Airways** (WT)

Nigeria Airways flies scheduled passenger services along an expanding route system that covers an intensive network: West and Central Africa, East Africa, the Middle East, Europe and the USA. The company was established in 1958 to take over the Nigerian operations of West African Airways Corporation (WAAC). The government assumed full ownership of the carrier and established it as the national airline on 1 May 1959. The airline operates a fleet of DC-10, Boeing 707, 727, 737 and A310-200 aircraft. Nigerian Airways also leases additional aircraft as needed. Its A310s can be found flying on routes that include Lagos-London (Heathrow) and Lagos-Port Harcourt-Amsterdam. *(Airbus Industrie)*

Opposite: **Olympic Airways** (OA)
Olympic Airways, the government-owned Greek national airline, was formed in April 1957 when Aristotle Onassis took over TAE Greek National Airlines. The initial fleet was composed of 14 DC-3s and a DC-4. The company maintains scheduled flights along an extensive domestic network and over international routes within Europe and the Middle East, Africa, southern Asia, Australia and North America. Olympic has a subsidiary company, Olympic Aviation, which operates general charter services throughout Greece and also operates scheduled local flights on behalf of its parent company. Olympic Airways maintains a fleet of Boeing 727, 737, 747, Shorts 330 and Airbus A300B4 aircraft. An example of the latter type can be seen here at London (Heathrow) after arrival from Athens. *(J Page)*

Above: **Philippine Airlines** (PR)
Philippine Airlines was established in 1941 and took over the routes of Philippine Aerial Taxi. The company maintains scheduled flights along domestic routes to over 40 points in the Philippines, as well as over international sectors to points in Asia, Australia, the Middle East, Europe and the United States. PAL was nationalised in October 1977 when the Rubican Corporation relinquished majority control to the state-owned Government Service Insurance System. The company operates a fleet of Boeing 747, BAe One-Eleven, 748, DC-10 and A300B4 aircraft. An example of the latter can be seen here at Hong Kong's Kai Tak Airport. *(R H Vandervord)*

Sabena — Belgian World Airways (SN)
Sabena, one of the world's oldest airlines, was formed in May 1923, as a successor to SNETA, Belgium's first air company. The carrier's inaugural service was made over a Brussels-Strasbourg route on 1 April 1924. The airline operates services to southern and eastern Asia, North Africa and the Middle East, Europe and North America. Sabena has a modern fleet of Boeing 737, 747, DC-10 and A310-200 aircraft types. The latter operate routes that include Brussels-London (Heathrow), Brussels-Larnaca-Tehran, and Brussels-Cairo-Jeddah. *(Airbus Industrie)*

Scanair (DK)

Scanair was established in 1961 by SAS, ABA Swedish Airlines, DDL Danish Airlines and DNL Norwegian Air Lines. The company is a prominent European charter carrier owned by Swedish, Danish and Norwegian interests. Scanair's main charter activity is the transportation of tour groups to and from the holiday resorts of several continents. Major destinations are mainland Spain, the Balearic Islands, the Canary Islands, Greece and Cyprus. Other important routes serve the UK, Austria, Gambia, Morocco, the USA and Canada. The airline operates a fleet of DC-8 and Airbus A300B4 aircraft. Other types are leased from both SAS and Linjeflyg as required. (R H Vandervord)

SAS — Scandinavian Airlines System
(SK)

Scandinavian Airlines System was formed on 1 August 1946 and was established to take over certain long-haul routes of DDL Danish Air Lines, DNL Norwegian Air Lines, ABA Swedish Air Lines and SILA Swedish Intercontinental Air Lines, the member carriers. A month after formation SAS assumed SILA's DC-4 service to New York, and extended routes to Brazil and Uruguay in the same year. The carrier has developed one of the world's largest service networks, linking over 80 points in Scandinavia, Europe, Africa, the Middle East, southern and eastern Asia, North and South America. As the national carrier of Denmark, Norway and Sweden, SAS has bases in all three countries. The airline operates a fleet of DC-8, DC-9, DC-10, MD-80 and F-27 aircraft. SAS no longer operates the A300 on major routes linking Scandinavia with European capitals as the aircraft capacity proved to be too great. *(A Clancey)*

Singapore Airlines (SQ)

Singapore Airlines operates its advanced fleet of A310 aircraft from its base at Changi International Airport. The type, along with the company's other regional aircraft, are used to 'fan out' passengers to destinations off the airline's main international routes from London, the USA, Australia etc. Singapore Airlines has had previous dealings with Airbus Industrie since the carrier has also operated the larger A300 type. The A310 can be seen at destinations including Hong Kong, Bandar Seri Begawan, Taipei and Jakarta. *(Airbus Industrie)*

43

South African Airways (SA)

South African Airways, the South African national flag carrier, was formed in February 1934. The company was a division of the South African Railways Administration and designed to take over from the faltering Union Airways. Following recommencement of services after the Second World War, SAA experienced a dramatic expansion period. The airline became jointly involved with BOAC and Qantas in serving London (Heathrow) and Perth respectively. Major jet types introduced included the Boeing 707 and 747 aircraft. The company is continental Africa's largest airline. With a major hub at Jan Smuts Airport in Johannesburg, South African Airways operates routes to North and South America, the Middle East, Europe, Australia, eastern Asia and the Indian Ocean, as well as to various regional southern African destinations. The airline currently operates a fleet of Boeing 737, 747 and Airbus A300 types. An example of the latter can be seen here on its delivery flight. *(Airbus Industrie)*

Swissair (SR)

Swissair, the highly acclaimed Swiss national airline, was established in March 1931 through a merger of Ad Astro Aero and Basle Air Transport. The company was designated as national flag carrier in 1947 and from then on a rapid growth in scheduled operations followed. Swissair is primarily privately-owned (77.8 per cent) with public institutions holding the remaining shares. The airline has, itself, a 57 per cent interest in Balair as well as a 57 per cent share holding in CTA, two Swiss charter companies. The airline maintains scheduled jet services to over 100 points along a vast intercontinental network that covers Europe, Africa, the Middle East, Asia, as well as North and South America. The company owns a fleet of Boeing 747, DC-9, MD-80, DC-10 and Airbus A310-200 and 300 types. An example of the long range A310-300 can be seen here.

(Airbus Industrie)

45

Above: **Thai Airways** (TH)

Thai Airways, the domestic and regional government-owned carrier, was formed on 1 November 1951 through a merger of Pacific Overseas Airlines (Siam), POAS and Siamese Airways Company — SAC. In August 1959 the airline and SAS reached an agreement to establish Thai Airways International. Thai Airways operates scheduled passenger services over a well-structured domestic route system and also serves regional international lines from Thailand to Laos and Malaysia to Viet-Nam.

The airline operates a fleet of Boeing 737, Shorts 330, 360 and Airbus A310-300 aircraft.
(Udo and Birgit Schaefer Collection)

Opposite: **Thai Airways International** (TG)

Thai Airways International operates Thailand's prime international services, connecting over 30 points in southern and eastern Asia, Australia, the Middle East, Europe and the USA. The company was formed in December 1959 by Thai Airways Company and Scandinavian Airlines System — SAS. Operations commenced in 1960, initially flying regional services in eastern Asia with Douglas DC-6B aircraft. The airline operates an all-jet fleet that includes Boeing 747, Douglas DC-10 and Airbus A300 aircraft. Both B4 and Dash 600 versions of the latter are utilised. Typical routes flown by these aircraft include Bangkok-Hong Kong, and Bangkok-Singapore. An A300-600 is seen here on its delivery flight. *(Airbus Industrie)*

THY — Turk Hava Yollari (TK)

THY, the government-controlled Turkish national airline, commenced operations in 1933 under the banner of Devlet Hava Yollari (State Airlines). The present name was adopted in March 1956, with the government owning a 99.9 per cent shareholding in the carrier. THY provides scheduled services with an all-jet fleet to various destinations throughout Europe, North Africa, the Middle East and Pakistan.

The company has a subsidiary, Cyprus Turkish Airlines, to which it leases various aircraft on an ad-hoc basis. In 1985 the company ordered seven Airbus A310-200s to replace its ageing DC-9 and Boeing 727 aircraft. The Airbus types are now used on routes from Istanbul to such destinations as London (Heathrow), Amsterdam, Frankfurt and Paris. *(Udo and Birgit Schaefer Collection)*

Toa Domestic Airlines (JD)

Toa Domestic was formed in May 1971 through a merger of Japan Domestic Airlines and Toa Airways. The company is Japan's second largest internal airline and schedules well over 300 daily flight departures from 38 points throughout Japan and the surrounding islands. Ownership of the airline is held by the Tokyo Express Electric Railway Company (26.23 per cent), Toa Kosan Ltd (10.57 per cent), JAL (9.19 per cent), Kinki Nippon Railway Company (8.72 per cent) and other interests (45.29 per cent). TDA itself has a 60 per cent holding in Japan Air Commuter, a carrier that serves local routes in the Nansei Islands. Destinations served by TDA include Tokyo, Osaka, Okayama, as well as many other Japanese towns. Primary traffic hubs on the TDA network are Tokyo, Osaka, Sapporo, Fukuoka and Kagoshima. The airline operates a fleet of MD-80, DC-9, YS-11 and A300B2 aircraft. The latter types are used mainly on lines from Tokyo to Fukuoka, Kagoshima, Kumamoto, Misuawa, Nagasaki, Ota, Osaka, Sapporo and Asakigawa. (Airbus Industrie)

Below: **Trans European Airways** (HE)
TEA was formed at Brussels in December 1970, with revenue flights beginning six months later. The airline operates worldwide passenger group charters and also provides leasing and contract services for other carriers. Maintenance activities are also performed through a division known as TEAMCO (Trans European Airways Maintenance Company). TEA passenger charter activities primarily involve package tour flights to and from European and Mediterranean holiday areas. Off season services include regular Moslem pilgrimage (Hadj) flights to Jeddah/Mecca. The airline operates a fleet of Boeing 737 aircraft as well as a single Airbus A300B1, OO-TEF, seen here at Athens. *(R H Vandervord)*

Opposite: **Tunis Air** (TU)
Tunis Air, the government-controlled Tunisian national airline, was created in 1948 by Air France and the Tunisian government. Initial operations were made with DC-3 aircraft and connected Tunisia with France. In the late 1950s lines were extended in North Africa and Rome and during the 1960s and 1970s a major route network was developed and a modern jet fleet assembled. Caravelle equipment was introduced in 1961, Boeing 727s in 1972, Boeing 737s in 1980, and a single A300B4 in 1982. The airline serves points within Europe, the Middle East, North and West Africa. Tunis Air operates domestic schedules to five points within Tunisia. *(Airbus Industrie)*

VARIG Brazilian Airlines (RG)

VARIG, Brazil's national flag carrier, was formed in May 1927, with the assistance of Kondor Syndikat. An inaugural flight over a Porto Alegre-Pelatos-Rio Grande route began the following month. Initial services concentrated in south-eastern Brazil, with an international line to Montevideo being added in 1942. From the purchase of Aérea General in 1951 the carrier extended routes as far north as Natal along Brazil's Atlantic seaboard. Intercontinental services began in July 1955 with the opening of a route to New York. With the acquisition of the REAL Aerovias consortium in 1961, VARIG expanded lines to Miami, Los Angeles, Mexico and Central America. Following the demise of Panair do Brasil in 1965, the airline took up routes to Europe. Today VARIG is the largest carrier in South America and operates a fleet of A300B4, Boeing 707, 727, 737, 747, 767, Lockheed Electra and Douglas DC-10 aircraft. An A300B4 is seen here at Toulouse, just prior to delivery.
(Udo and Birgit Schaefer Collection)

VASP Brazilian Airlines (VP)

VASP was formed in November 1933 by the São Paulo State government, the city of São Paulo and the municipal banks. Operations began using a three-passenger Monospar linking São Paulo with San Carlos, Rio Preto, Ribeirão Prêto and Uberaba. In 1962, as part of the nationalisation in Brazil, VASP took over Loide Aéreo Nacional and Navegacão Aéro Brasileira (NAB). VASP operates a huge domestic network from Belem in the north to Porto Alegre, and from Recife on the east coast to Rio Branco on the Bolivian border. High frequency services are operated in conjunction with Cruzeiro, Transbrasil and VARIG over the Rio-São Paulo and Rio-Belo Horizonte-Brasilia routes. These are better known as the Ponta Aérea (Air Bridge). VASP has operated three A300B2s since 1982 and uses them to serve such destinations as the freeport of Manaus, 1000 miles from the Amazon Delta. A300B2, PP-SNN, is seen at one of VASP's destinations, patiently awaiting another load of passengers. *(Udo and Birgit Schaefer Collection)*

53

Below: **Airbus Industrie A300**
The first of the Airbus family, the A300 entered service in May 1974. A wide-bodied, twin-aisled airliner designated for short to medium range service, the A300 was the first twin-aisle twin offering passengers new standards of comfort and operators outstanding performance in reliability. The Airbus partnership is made up of Aérospatiale SNI (France), Deutsche Airbus GmbH (West Germany), British Aerospace (UK), Construcciones Aeronauticas SA (Spain), Fokker NV (Netherlands) and Belairbus (Belgium) as associate members. The latest member of the Airbus fleet, the A300 'Fly-By-Wire', made its public debut at the 1986 Farnborough Air Show in the UK. The aircraft is the latest in 'state-of-the-art' flying. *(G Maxted)*

Opposite: **Airbus Industrie A310**
The A310 entered service in March 1983 to meet the requirements of the most difficult air routes. Shorter than the A300-600, but with the same fuselage cross-section, it incorporates all the technological advances made by Airbus Industrie over the last ten years in aerodynamics and on-board electronic systems management. The A310 has demonstrated unequalled reliability and the most recent version, the A310-300, is capable of ranges of up to 9300 km (5000 n miles). *(Airbus Industrie)*

Air France (AF)

Air France operates its A300s on, among others, one of the world's busiest air routes, that from Paris (CDG) to London (Heathrow). The carrier provides scheduled services over one of the world's biggest and best established intercontinental route networks, stretching over 60 000 km (37 300 miles) around the globe. The government-controlled French national flag carrier links France with over 130 foreign points, as well as serving several domestic trunk routes in France itself. An A300B4 is seen here while on finals for London (Heathrow). *(A Clancey)*

Air-India (AI)

Air-India, the Indian flag carrier, was formed as Tata Airlines in October 1932. An inaugural service was flown over a Karachi-Ahmedabad-Bombay-Bellany-Madras route. It was not until 1946 that privately-owned Tata became publicly-owned Air-India. The company has subsidiaries that include Air-India Charters and the Hotel Corporation of India. The carrier also has holdings in the Indian regional airline, Vayudoot. 1986 saw the delivery of Air-India's first A310s. The aircraft, along with their older and larger brothers, the A300s, are used on routes from Bombay to the Arabian Gulf and other regional destinations. An A310 is seen here in the full livery of Air-India just prior to delivery.
(Udo and Birgit Schaefer Collection)

Opposite: **Condor Flugdienst GmbH** (DF)
Condor was formed by Lufthansa and other interests in December 1955 as Deutsche Flugdienst. Flight activities began in the spring of 1956 with Vickers Viking aircraft. In 1960 the carrier became fully owned by Lufthansa and the following year changed to its present name. Condor is one of the world's largest charter airlines. Intercontinental tour and general passenger flights are carried out by an all-jet fleet. Regular tour charter services operate from four primary West German airports, at Frankfurt, Düsseldorf, Munich and Stuttgart. Condor operates package tour flights to Europe, North

and East Africa, the Far East, as well as to North and South America. The airline has a fleet of Boeing 727, 737, DC-8, DC-10 and A310-200 aircraft. An example of the latter can be seen here on its delivery flight. *(Airbus Industrie)*

Above: **Korean Air Lines** (KE)
Korean Air Lines was formed on 1 March 1969 with an inaugural service over a Seoul-Osaka-Taipei-Hong Kong-Bangkok route in October of the same year. In 1971 KAL opened a transpacific cargo service to the USA, connecting Seoul with Los Angeles by way of Tokyo and Anchorage. Two

years later the company took possession of its first Boeing 747, which went into immediate service on this route. In 1975 Korean Air Lines took delivery of three Airbus A300B4s and used them on medium-range regional routes. 1981 marked a significant turning point for the airline when it became the first East Asian carrier to operate passenger services between Seoul and Tripoli (Libya). When Korean Air Lines changed its name to Korean Air in 1984, the company was operating a fleet dominated by Boeing types and flew from its base in Seoul to the USA, Europe, the Middle and Far East, and North Africa. *(Airbus Industrie)*

59

Kuwait Airways Corporation (KU)

Kuwait Airways was formed in March 1954 as Kuwait National Airways Company and operated its first service over a Kuwait-Basrah route two months later using DC-3 aircraft. By mid-1955 the Kuwaiti government had gained a 50 per cent holding in the carrier, which then adopted its present title. Full government control was realised in 1962. Two years later KAC inaugurated Comet jet flights to London, Paris, Frankfurt and Geneva, and shortly afterwards acquired Trans Arabian Airways Kuwait. The company operates services to New York, Europe, southern and eastern Asia, North Africa and the Middle East. The airline has a modern fleet of A300-600, A310, Boeing 707, 727, 747, Gulfstream 3 and BAe 125 aircraft. An A300C-600, 9K-AHG, is seen here at Amsterdam Schiphol Airport on a freight service. *(Airbus Industrie)*

Lufthansa German Airlines (LH)

Lufthansa was established in 1926 when Deutsche Luft Hansa was formed through the merger of Aero Lloyd and Junkers Luftverkehr. Following the Second World War, the company was dissolved. A new West German national airline known as Luftag was established in 1953, being renamed Deutsche Lufthansa a year later. Lufthansa subsidiaries include the charter airline Condor Flugdienst and the freight airline German Cargo Services. Financial interest is also held in DLT German Commuter Airlines. The company operates a fleet of Boeing 727, 737, 747, DC-10, Airbus A300 and A310 aircraft. Lufthansa has an order for the A320. An A300 is pictured here awaiting passengers.
(Udo and Birgit Schaefer Collection)

61

Pan American World Airways (PA)

Pan Am, arguably the pre-eminent US international airline, provides scheduled jet services over a 400 000 km (250 000 mile) route network, linking over 90 points around the globe. The company has its world headquarters in New York and has major bases at Miami, New York (JFK), Los Angeles, San Francisco, London, Frankfurt, West Berlin and Tokyo. The A310 aircraft are used on Pam Am's internal West German services, as well as for transatlantic crossings via London (Heathrow). The A310-200 types currently in service will be eventually replaced by A310-300 and A320 aircraft. *(C Huybrechts)*

Above: **Malaysian Airline System** (MH)
MAS was formed in April 1971 to succeed Malaysia-Singapore Airlines. The Malaysian national flag carrier is owned by the government (70 per cent) and private shareholders (30 per cent). Operations started on 9 June of that year over a local network that linked Singapore and Kuala Lumpur with Ipoh, Penang, Kuantan and Kota Bharu. The airline changed its name to Malaysia Airways in November 1963. The carrier flies a large domestic network as well as long-haul services to Europe, the Middle East, Australia, the Far East and the USA. A fleet of A300, Boeing 737, 747, DHC-6 Twin Otter, F-27 and DC-10 aircraft is utilised. *(A Clancey)*

Overleaf: **Singapore Airlines** (SQ)
Singapore Airlines, the ever expanding Singaporean flag carrier, was formed in 1972. This followed the dissolution of the bi-national Malaysia-Singapore Airlines (MSA). SIA maintains scheduled flights over a vast route system that links over 40 points in southern and eastern Asia, the South Pacific, the Middle East, Europe and North America. The company is organised as a subsidiary of Temasek Holdings (Private) Ltd, with ownership controlled by the government (82.85 per cent) and carrier employees (17.15 per cent). SIA subsidiaries include the air carrier Tradewinds Ltd as well as airport services, hotel maintenance, real estate and insurance enterprises. The current SIA fleet comprises Boeing 747, 757 and Airbus A310 aircraft. The company's A300 aircraft, one of which is illustrated here, have now been withdrawn from service and have been sold. *(Airbus Industrie)*